AMONG THE PAPER DAISIES

Writings on ecology and being

Microcarpa
PRESS

First published 2026

Microcarpa Press
PO Box 950, Castlemaine
Victoria 3450, Australia

Cover by Kalm Designs

ISBN: 978-0-646-73852-9

A catalogue record for this book is available
from the National Library of Australia

For all the many beings

TABLE OF CONTENTS

PRELUDE

Recollections of place-making

I scrambled up the tree, my feet loosening the flaky bark as I sought footing, and climbed into the make-shift tree house. It was nothing more than an old shipping pallet in the fork of a grey eucalpyt tree, with a hand-painted sign that read 'NO BOYS ALLOWED'—once I saw boys stop on their bikes and throw stones at it.

Making my way out onto a branch, I straddled the tree, legs dangling either side. The sun felt warm in my hair and I looked out towards the dirt road. My legs swung restlessly as I waited impatiently for my friend to arrive.

∞

On all fours, we crawled through the maze-like thicket of Drooping Cassinia. I followed my friend through the scrub;

this way, that way and around, until it opened into a little clearing.

'Look!' she pointed.

And there it was. A great big stone in the middle of the bush, with a flat top. The perfect seat for two. So we sat together and shared our picnic within the scrubby walls of our own secret garden.

∞

Mum excitedly hurried me outside, showing me to a spot in the understorey.

'An Echidna!' It had curled up and buried itself into the leaf litter. 'Would you like to touch it?'

I looked at the blonde-tipped spikes with dark roots. 'Won't it hurt?'

'Not like this', she said. Very gently, she stroked the spikes from front to back, like we pat our dog without ruffling his fur.

I reached out and tenderly touched its smooth spikes. A thrill leaped through my body as I felt its own little body wriggle slightly under my fingertips, burying itself deeper into the leaf litter.

Then we quietly said goodbye, leaving the Echidna to emerge safely and carry on foraging for ants.

∞

I remember another time, when I came home on my bike from school, Mum met me on the driveway. 'I found some Nodding Greenhoods!'

I followed her, and we squatted down as she pointed out the delicate orchids. They were like green-capped pixies, clustered together among the undergrowth, with droplets of winter rain dripping from their chins.

From winter through spring and early summer, we watched the wildflowers and orchids emerge in succession. Each one was named to me, with their perfect-sounding names†, like the fairies in my picture books. Early Nancy, Scented Sundews, Waxlip Orchids, Pink Fingers,

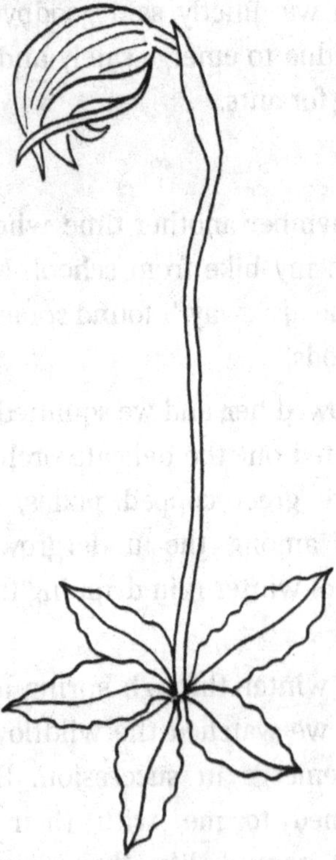

Milkmaids, Slender Sun Orchids, Fairy Wax-flowers, Chocolate Lilies.

Until eventually, the bush was grey, the dam was dry and cracked, and the sun scorched my skin. I noticed only the golden Sticky Everlastings, with their fluffy seeds bursting from their tufted centres and being carried away by the hot wind.

I remember wandering and playing among these paper daisies. There was so many that Mum once dried a bunch, and placed them in a vase on the bathroom counter. And they stood there—*everlasting*—becoming part of the house, as spiders made cobwebs in their yellow cornices.

I can still see the bouquet in my mind's eye. Static. Watching my strange existence, instead of dying back in the summer time, to the thrum of cicadas.

LEARNING FROM BATS

Bat Cave: with nature

That familiar earthy smell and moment of cool as I place my feet on the ground. 'Off belay', I call up to the surface, so they know that I am safe, and unclip the carabiner from my harness. The belayer at the end of the rope waits in the sun above as I head off down the rock-fern dotted rock pile; into the outstretched arm of the natural amphitheatre; into the darkness.‡

With each step, I'm keeping watch for snakes that may have fallen into the cave, with no way out; sunning themselves at the bottom, where the collapsed 'roof window' entrance brings in some sunlight. It is the same way that many animals have fallen to their end in the caves to become the world heritage listed fossil deposits that they are known for today at

Naracoorte Caves, South Australia. From small critters, through to Thylacoleo (the marsupial lion) and Diprotodon (a wombat the size of an elephant) that once used to walk these plains.[1]

Today, it is 37°C outside. In the cave it is a pleasant 20 degrees or so. And when I return here in the winter when it's just 12 degrees outside, I'll drop into the cave and take off my jacket because it will *still* be a pleasant and stable 20 degrees.

I walk into the cave with my headlamp on, which is covered in red cellophane to filter the light. And just beyond the twilight zone I stop for a moment to let my eyes adjust to the darkness. I turn down my head torch to just the faintest red light, seeing barely a metre ahead. I know the way well around the rocks and through the blackness.

I can hear them fluttering past my helmet. There is constant movement of bats. It is a busy thoroughfare; the only link between the two busiest chambers (or rooms, if you like) of the cave. The large 'maternity chamber' where the majority of the population gathers and the mothers birth and raise their young, and an adjoining section of the cave, the 'drinking chamber'—where they drink from a large stalactite dripping with water slowly seeping through the limestone.[2]

The cave system is huge. My research equipment is further than 100 metres in the cave, monitoring bats that fly past. I always feel so small here. For such small bats (weighing just 15 grams) they roost in enormous caves. The cave ceiling is high and the darkness feels all-encompassing. It is at this point that I like to take a moment to remember that the bats have been flying in this passage, as they are now, for at least half a million years. We know this because of the fossil deposits in the cave. And it makes me feel even smaller. A tiny blip. A mere comma in a much bigger story.

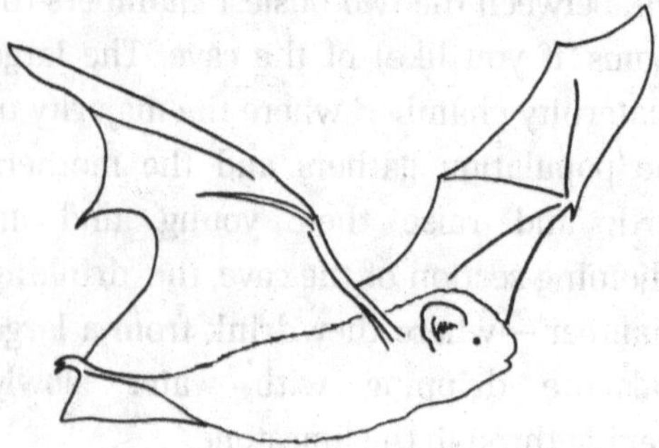

It feels so quiet and contemplative. A meditation perhaps. But then that space I've created in my heart is filled with grief and disbelief that these bats that still in the 1950s were in abundant numbers, are now critically endangered. It's difficult to grasp sometimes. I close my eyes for a few seconds and put all my heart and mind into my own will, pushing my own hopes for the story. But I really don't know what futures will be written for these little beings.

And I suppose that is where this story begins. Bats. And time.

What even is time? It *bends*. It's *relative*. And the more you read about it (well, at least when you're not a physicist), the more your brain boggles.

Ambelin Kwaymullina writes that linear time is something that Europeans brought to this continent:[3]

> 'In Indigenous systems
>
> time is not linear
>
> It moves in cycles
>
> It exists in space
>
> in Country
>
> and is as susceptible
>
> to action and interaction
>
> as any other life'

This type of understanding helps me breathe more easefully, instead of feeling like I have fallen into a deep empty chasm. Because my human brain has a difficult time trying to imagine half a million years in the linear framework I was given. There are timescales my human mind can't fathom.

These bats, now called Southern Bent-wing Bats, have survived ice ages. They have survived the ocean rising and flooding inland and then receding, many times over.

These bats live *with* nature, rather than trying to fight its tides. One way this species (and others) have done this is by inhabiting incredibly stable cave environments. While many bat species roost in tree hollows, or human-made structures, those that live in caves can escape extreme temperature shifts, and use particular microclimates in the caves to their advantage, much the same way that gardeners will take advantage of different microclimates in their garden for different plants or seasons.

It was one of the first personal lessons I reflected on when I began researching bats, which I was reminded of every time

I collected data over the course of my PhD and dropped into that cave at a comfortable 20 degrees. I try to remember to continue to ask the question, how can I/we live *with* nature? How can I live with the seasons? How can I live with all its extremes or threats, without expending too much energy? How can we do this better?

∞

To live *with* nature means to *know* nature. And surely, to know nature is to know ourselves, since we are, and are a part of, nature.

There are endless ways to get to know nature. Though I believe it usually requires *slowing down*.

Noticing. Observing. Deep listening. To name a few.

I remember as a first year uni student, creating my first nature diary for naturalist studies. It required making observations, identifying species and processes, and drawing. And to draw, you needed to *really* look, and look closely. Otherwise, you'd find yourself drawing what you *thought* you knew, which was in fact often not what was really there.

I remember sitting by a Sticky Everlasting thinking I'd make a quick nature diary entry. It was one of my favourite wildflowers which I thought I knew well, and one which provides me with a great sense of place, because it grew in abundance around the bush block where I grew up here on Djaara country. I sat cross-legged on the dusty and rocky ground, with my A4 sketchbook and pencil. When I started to look, I saw more and more. Such a diversity of insects coming and going, visiting its offerings of nectar and pollen, and walking up and down its sticky green leaves and stems. And then to my delight and surprise, I noticed a yellow flower spider, camouflaged under the yellow papery bracts of the flower, waiting to pounce on an unsuspecting visitor.

By getting to know the Sticky Everlasting, I was coming to understand more than just its form and colour, and my childhood memories. I was getting to know its many relationships within our ecosystem, and by doing so, I was getting to know so many other critters and aspects of place.

Seasons and wintering

In the winter, Southern Bent-wing Bats become less active. The temperatures are cool and insect prey are less abundant. Bats will use a process called torpor. They will find suitable caves or microclimates, and lower their metabolic rate so that their body temperature is only ever so slightly above ambient temperature. Depending on the conditions, they do this for a period of hours, days or weeks at a time, surviving on fat stores accumulated in the months prior.[4]

For Southern Bent-wing Bats, mating occurs in autumn, but females will delay implantation of the fertilised egg and pregnancy to allow for winter torpor,[5] and to time the birth of their pup for over the warmer months when insects are more plentiful. Other bat species will do this by delayed fertilisation—the females store the sperm over winter and 'self-impregnate' themselves come spring!

Bats remind me to take time to rest. To 'winter', as a verb. In *Wintering*, Katherine May observes, 'Plants and animals don't fight the winter; they don't pretend it's not happening and attempt to carry on living the same lives they lived in the summer'.[6]

Nature is cyclical. The Earth's rotations and therefore the sun's seasons, the moon's phases, the daily cycles of day and night. The cycles of life and death as we are all made and destroyed or decay. Things are always changing. Always shifting. By understanding and living with the seasons, I can see that everything (no matter how difficult) is temporary.

In nature, there is a time for everything. A time for foraging, hunting, growing. A time for mating, nesting or flowering. Times for resting, for conserving energy and resources. Nature and its many ecosystems are complex. These times will vary between organisms or beings, and climates, and will be influenced by many things.

How the seasons are identified varies the world over, between indigenous peoples and colonisers, and between different climatic zones and landscapes. When do the seasons change really? Is it when a particular flower blooms, or another? When the rain begins? When the rain settles in? Any moment could be a new season.

There is no right or wrong to consider or demarcate the seasons and cycles, though much can be learnt from First Nations calendars where this cultural knowledge has been shared. And I encourage questioning the seasons that have been taught, to see if they make sense where you live, and to you and your lived experience.

Where I live in central Victoria, the winters are very cold, frosty, and also when we receive most of our rainfall for the year in this dry, inland ecosystem. With the rainfall bringing much-needed moisture into the soil, winter means abundance for some species here. Many of the eucalypt trees here bloom in the cooler months, attracting birds from great distances. This is also an important food source for flying-foxes (yes, flying-foxes are bats!), and in turn, flying-foxes are important pollinators because they fly large distances each night —generally much further than insects, birds and small-mammal pollinators.

The eucalypts do not flower all at once. There is a time for one species, and a time for another, or for one tree or another; just like the successive flowering of orchids and wildflowers in this landscape come spring. Within even a single species of tree, not everyone needs to flower at once, *or even at all,* each season.

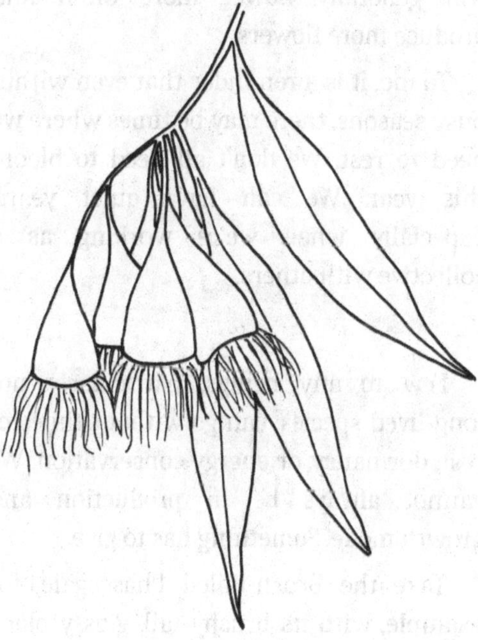

Flowering and nectar in these forests is said to be 'patchy' over space and time.[7] Mobile species will follow and seek out this flowering. And larger, more mature trees, will generally flower more often and produce more flowers.

To me, it is a reminder that even within busy seasons, there may be times where we need to rest. We don't all need to bloom this year. We can have quiet years. Especially when we're working as a collective with others.

∞

Few to any beings, particularly not long-lived species, can go without times of rest, dormancy, or energy conservation. We cannot always be in production and growth mode. Something has to give.

Take the Brush-tailed Phascogale, for example, with its brushy tail, glossy black eyes, and 'racing stripe' down its snout. They are agile creatures, foraging and hunting up and down tree trunks and within the treetops. Feeding on diverse prey sources, from insects to blossoms, to small reptiles, nestlings, and occasionally larger prey.

Phascogales have an unusual life cycle.[8] When males disperse from their birth territories for the breeding season, they enter a period where they do little else than focusing on finding mates, searching up to several kilometres each night. Following this mating frenzy, all males die. They are weakened and preyed on, or succumb to disease or stress-induced illnesses. Even most females only usually live to raise one litter.

This life cycle strategy has pros and cons. It's risky, but frees up competition for resources between generations and for the females to raise their young. This strategy is only effective for one that is short-lived, similar to annual plants that live a full-but-short life, relying on their seeds for the next generation.

These approaches to life are not suitable for long-lived species that need to survive and thrive in all seasons, different stages of life and conditions, dry years and wet. And I think, we therefore shouldn't be

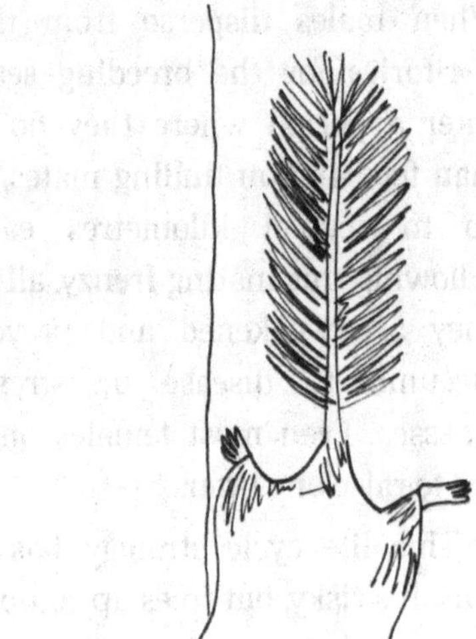

expecting of ourselves that we can manage to do all the things and be productive at all times either.

In a modern world that does not value and prioritise rest, or provide access to this basic need equitably, taking rest can itself be a political action. In *Rest is Resistance*, Tricia Hersey says, 'I wish you a deep knowing that exhaustion is not a normal way of living. You are enough. You can rest.'[9]

Bats teach us to rest and 'winter' in whatever way makes sense for each of us and the place that we live. Rest or refuge is not just needed from the cold or dark, it can also be needed from hot periods or dry seasons, or from too many social gatherings where we may need some time of comparative solitude. The use of torpor in bats was long considered a cold climate adaptation, but it has since been shown that it used by bats the world over, in tropical climates to temperate, and in all seasons, just to different degrees, patterns or times.[10]

If we don't rest, not only do we risk burning ourselves out, but we also risk making too many mistakes by pushing in the wrong directions. Wakaya man Yin Paradies shares that, 'in times of urgency it is best to slow down, so that we are not lured into reproducing the same responses that got us into the mess in the first place'.[11] This requires time to think and reflect—something that comes more naturally in times of rest or retreat. Times of *not* doing.

On hope: think local, act local

In spring, Southern Bent-wing Bats start congregating at maternity caves. Females and males, adults of all ages and young from the last breeding season.[12] By coming together *en masse* and due to the domed ceiling of the maternity chamber in the cave, the heat of the bats themselves and their activity raises the temperature of the cave by an *additional* 10–12°C.[13] In this way, the maternity chamber is warm and humid, and acts like a humidi-crib, facilitating the survival and growth of the pups, which are born furless.

These conditions also allow the lactating females to be able to leave their pups clustered together on the cave ceiling in a 'creche', so that they can forage long distances from the cave to search for food without needing to return regularly through the night to suckle their young.[14] And amazingly, when the mother returns, she knows exactly which pup is hers among the hundreds or thousands of pink pups.

∞

There is much conversation on the role of hope in environmentalism. There has been criticism from within the field that ecologists are too pessimistic, and the need for hope as the 'elixir of action'.[15]

In an interview with climate change ecologist Lesley Hughes, she says, 'Hope I regard as a strategy rather than an emotion… for me, remaining hopeful is quite a conscious decision'.[16] I too have regularly used hope as a strategy to face the day.

But what is hope really? I will try to dance around this topic carefully, because I am an ecologist not a psychologist. But, as it turns out, hope is rather complex.

It 'is a multi-faceted emotional state or motivational attitude with many subtypes, including authentic (or active), passive, false, and radical hopes'.[17] Some of these, as you will have guessed, are more helpful for environmental action than others.

Grief and hopelessness are described as being a common emotional load in conservation and environmental work. In psychology, hope is often said to be based on a combination of having a specific objective, possibility of success (even if it is a low probability) and personal agency.[17] When considering these factors, I can't help but feel like it's no wonder so many of us (both within the environmental sector and more broadly) feel hopeless, or are having to actively seek or cultivate hope. Objectives can feel huge, complex and far-reaching. Success can feel improbable because of the enormity of it all. And there can be a lack of a sense of empowerment or ability to control anything outside of our sphere of influence.

For years, my partner and I have debated on the importance and value of hope in our present times. Rather like the characters Andy and Red in the movie *The Shawshank Redemption*,[18] one of us tends to declare hope to be a good thing, and the other feels that it can be dangerous. But as time has passed, I'm starting to feel like hope, for me, is neither here nor there. And rather than focusing on cultivating hope, I'm beginning to focus more and more on where I can *act*, especially with others in my community.

Things tend to make more sense on a local level. Humans are community animals, dependent on the many relationships within their ecosystems. On this level people can work together in a way that understands and works *with* ecologies and environments, strengths and deficiencies, cultures and values. Through collective action (and rest) on this scale, we can make things happen.

For millennia, Southern Bent-wing Bats have congregated together in large numbers at maternity caves. Through collective action on a localised level, they create change. They create a real and measurable outcome. They create the conditions they need to thrive. The ingredients here appear to be collective action on a local scale. These bats couldn't achieve this outside the cave.

I see this happening in my own community. My local 'Friends of' group has just celebrated 25 years of action; rehabilitating the local creek that had been completely cleared and turned over by colonisers in search of gold. Today the creek has mature trees, mid- and lower-storey plants, and diverse wildlife including Platypus. And as a direct result of community action on a similar time-frame, the creek now has more than five kilometres of meandering walking and cycling track connecting to town, without needing to cross roads, which means that lots of people are walking and riding more for transport, exercise, or time in nature.

I suppose what I'm talking about here is the well-known motto of 'think global, act local'. Which may be true. *Maybe* the ripples from these actions will reach beyond. After all, everything is connected. And people are doing these types of things the world over. But I am just one small human. I need to think local and act local. That's where the magic is.

Bringing it all together

Bats are fascinating (as I'm sure you will agree). But this reflection could have been threaded together by anything really. In my view, it's the *ecology* that really matters here, because that is what we are part of. The many interwoven relationships between living beings and environments.

In his philosophical paper, 'What is it like to be a bat?' Thomas Nagel makes the point that we cannot know what it is like to be a bat (or anything else).[19] I certainly have no idea what it's like to be a bat, paper daisy, gum tree or phascogale. But observations of their ecology have given me a lot to reflect on. They have taught me about how to keep on *keeping on*, in a late capitalist world, in the face of climate change, extinctions and doom scrolling.

I have tried to become a 'noticer'. To observe the seasons and cycles and processes at play. To celebrate them wherever I can. To see how I can better live *with* nature and the place I am *of* and in. To rest and winter. And to think and act on this local level. A level that is place-based and community-based.

By bringing these together, I feel as though I am *becoming* a thread in the web. I say 'becoming', because it is a verb, a process, and something that feels intentional.

We are always part of the ecological web, but it is easy to feel detached, or keep trying to act outside of its bounds. It's a constant process of remembering. It is something I am constantly learning and re-learning. But each time I remember, I start to breathe more deeply, instead of feeling afraid or exhausted.

Slowly. Season by season.

EARTHSHINE

The moon in our sky

When one of my children was very small, they pondered while looking at the moon, 'Mummy, why does the moon fly in the sky like birds and aeroplanes?' I loved this question for so many reasons, and in part, simply because often the moon goes unnoticed and unquestioned.

I remember in primary school taking the red-bound children's encyclopedia off the shelf to write an assignment about the moon landing. Those of my parents' era share stories of their experiences and excitement of the moon landing as kids, huddling into neighbours' homes to watch it, because few people had televisions. 'One giant leap for mankind', but for all the fuss and 'progress', I don't recall any moment in my childhood being taught about the moon itself or its phases. I don't recall even really noticing the moon, let alone wondering about its influence on us and other beings on Earth. I suppose I just vaguely understood that it's there in the sky sometimes (and sometimes it isn't) and that it changes appearance.

We largely associate the moon with the night, but really it's equally visible in the daylight hours, if one went outside at the right time of day or night and watched it traverse the sky through the moon cycle.

The moon is visible from Earth in continuous, cyclical 'phases' based on the portion of moon illuminated by the sun. The most conspicuous phase is perhaps the full moon, which rises around sunset and sets around sunrise—especially during the shorter days of winter, when we are likely to see the full moon lifting into the sky as we may be returning home for the evening.[20]

From the full moon, the moon gradually 'wanes' over the subsequent fortnight, each evening rising later and with less of the moon being visible. The new (or dark) moon rises around sunrise and sets around sunset, and from here 'waxes' with increasingly more of the moon becoming visible as it rises later and later in the day, until it has reached full moon again.

I started observing the moon cycles in recent years. Just noticing them, like mini seasons, because it was something that had previously gone unobserved in my world.

I've grown to love noticing when the nights are filled with moonlight, and when they are dark. Learning when and where in the day the moon can be seen, and its constant changing appearance of the phases. Always changing, and always returning.

My favourite moon gazing is a crescent moon, when the full disc of the moon is still faintly visible from 'earthshine'—the light reflecting back from Earth onto the moon. It fills me with awe, and all I can do is stop and feel wonderfully small and unknowing.

Ecology of the moon

The moon is an important part of our ecology. The ocean's tides are caused by the gravitational pull of the moon; they shape our coastal ecosystems, with rock pools being exposed and dry, then thrashing under waves. Large 'spring tides' coincide with new and full moons because the sun is lined up with the Earth and moon, adding extra gravitational force.[21]

The changing light across the lunar cycle profoundly changes the experience of the night. From darkness, to being bathed in silver light. It can even *sound* different.

Among the songbirds that dominate the early phase of the dawn chorus, there appears to be different thresholds of light that will trigger each to sing, allowing for

their song to be spaced between species with less interference. Moonlight can cause Australian Magpies to start singing in the night, before twilight or the dawn chorus.[22]

Few frogs are unaffected by the moon. Most frog species either increase or decrease activity or reproductive behaviour around full moon, with responses being highly species-specific.[23] The main reasons for these responses appear to be predator avoidance and breeding synchronisation. This means that there are different frogs calling, and levels of call activity, depending on the moon cycle.

Some bats are said to be 'lunar-phobic', with activity decreasing with increasing moonlight at night during the lunar cycle.[24] Other bats either don't appear to be a significantly affected, or have instead been shown to switch foraging habitats during different luminations, which could be associated with availability or locations of insects.[25]

Even on the rainforest floor, where little light permeates through the trees, animal activity varies according to the moon.[26] Moonlight tends to suppress wildlife activity due to predation risk, though this differs between animal groups, habitat types and sensory abilities.[27]

Counter intuitively, increasing moonlight can actually make it harder for some predators to hunt, due to it being easier for prey to spot their predators. And perhaps even more counter intuitively, white Barn Owls (which you'd expect to be easy to spot) appear to have evolved a way to negate the effect of the lunar cycle on their hunting success. The white plumage of Barn Owls startles rodents by reflecting the moonlight, which results in prolonged 'freeze' times and therefore improved catchability. By comparison, Barn Owls with darker plumage are less successful hunters during full moon.[28]

The lunar cycle can also act as a synchroniser of activity. Once a year, when conditions are right, the corals of the Great Barrier Reef spawn directly following the full moon.[29] The phase of the moon is also associated with the timing of migration in some birds and insects.[29,30]

The rising moon has been shown to help moths find mates, and the position of

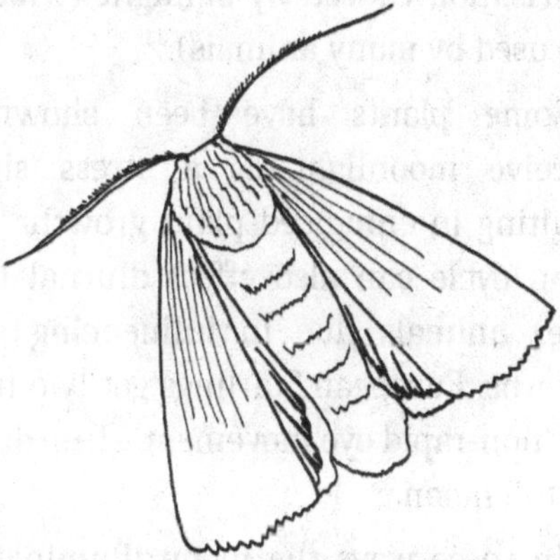

the moon helps them orientate.[31] The polarisation of moonlight (patterns caused by scattering of light when it hits tiny particles in the atmosphere) is used by dung beetles to orientate themselves, so that they can efficiently roll away their balls of dung in a straight line away from competitors.[32] Nocturnal bull ants use this moonlight polarisation to return home, despite this polarisation being a literal million times dimmer than the polarisation caused by sunlight (which is also used by many animals).[33]

Some plants have been shown to perceive moonlight as a stress signal, resulting in enhanced plant growth.[34] The moon cycle can also effect diurnal (day-time) animals due to influencing sleep patterns. European Starlings get two hours less 'non-rapid eye movement' sleep during the full moon.[35]

In these ways the moon illuminates a truth that *everything* is connected—just by its very nature of being. Even the moon, some 384,000 kilometres away from Earth.[36]

Inviting in the darkness

Once, as an undergraduate university student, we were tasked with mapping out a limestone cave. Clad in overalls, and topped with helmets and head torches, this involved exploring side tunnels and caverns, and taking measurements of distance and incline to contribute detail to the map.

I found myself crawling in the sandy floor of a small tunnel, away from other students, and I stopped, noticing the immense quiet. I turned off my headlamp and was in *complete* darkness. It was as if some kind of gentle instinct or intuition came over me, and I put my head to the side and lay there a minute.

It was one of the most peaceful moments I've ever experienced.

I felt *held*. Held in the quiet darkness within the Earth.

∞

Before artificial lighting, the moon cycle would have been a tangible presence in human lives, providing times of darkness, as well as soft night light. The moon influenced social lives and cultural events, hunting opportunities, and the more mundane such as sleep patterns (just as it does for starlings).

The moon cycle has been the basis of many calendars, and lunar farming (timing agricultural practices based on the moon cycle) is practised in various ways across different cultures. There are many beliefs and cultural traditions held to the moon the world over, though most of them are not my stories to share.

Where my ancestors lived in the low lands of western Europe, there are hints to be found about the importance once attributed to the moon. Writings exist which document a long sermon given by Eligius, a Christian missionary, preaching against the many pagan customs that were being practised there, including some connected with the moon. He taught that no meaning was associated with the first work of the day or phase of the moon, that no one should abandon work because of the new moon, or call the sun or moon 'Lord'.[37] This suggests that the moon was revered and respected, as well as being an important part of the calendar and rhythms of work and life.

One of the ways the moon may affect our lives is sleep.[38, 39] Living in the country, with dark starry skies and less street lighting, I sometimes notice more wakefulness around the full moon when the light creeps under the curtains.

A study undertaken in urban, regional and indigenous communities of Argentina, and in university students in the United States, showed a clear association of the moon cycle with duration of sleep and the timing of sleep onset.[39] Sleep was generally shorter and occurred later around the full moon. However, shortened sleep duration and delayed onset of sleep also occurred with increased access to artificial light.

Intriguingly, in another sleep study, participants that experienced long dark nights, removed from all sources of artificial light, began sleeping eight hours —but broken into two bouts, that were separated by 1–3 hours of quiet rest.[40] The results provide clues about what human sleep cycles may have been like prior to industrialisation, before lighting became common in our streets and homes. This is supported by historical research in western Europe, which found numerous records confirming that people used to have a 'first sleep' and 'second sleep'.[41] Between these sleeps, the intervening time of darkness (or low light) was used for activities such as prayer, contemplation and intimacy.

Through the solar and lunar cycles, darkness was once inherently present in the flow of our lives, like the spaces between the notes. For humans, darkness at night is the cocoon of rest, a remover of distractions, and may even be important for our health.[42,43]

I love natural light. In the morning, I open the curtains, and watch the sunlight glimmer on the foliage outside while sipping my coffee. Light is essential for my mental health.

But for me, following the moon cycle has invited in the periods of darkness, as much as the light. Life isn't a constant full moon, and nor do we want it to be. I have found that, like silence, darkness needn't be something for us to fear—it may even be precious, and it is certainly increasingly scarce.

Connecting in

Despite the moon being intertwined with ecology on Earth, it seems as though its influence is now beyond mainstream awareness, or has been relegated to myth and the unreal. But what even *is* real?

When I was in high school, I was told, 'Just wait 'til you get to the real world'—trying to instil fear in me. But I was already in that world.

Exams, grades, expectations, *fear*, bullying, stereotypes, judgement. Greed, isolation, shame, homophobia, racism, sexism, *fear*. Exponential growth, plastic, concrete, coal, petrol, supermarkets, *fear*. Production, fashion, success, failure, external validation of worth.

And then came social media, news, scrolling.

Separation. Alienation.

Me.

You.

Other.

In truth, so much of the 'real world' we are taught to fear is a human fabrication. A delusion, a disconnection; but with real consequences and real suffering.

The *real* world is the world in which all this is situated. The real world is the Earth, the rock, the sand, the soil, the roots, the plants, the fungi, the bacteria, the water, the wind, the animals, the sun, the sky, the stars… the moon.

One day I became distinctly aware of the fear underlying everything. It was under anything I had been conscious of, but was influencing all. Deep-seated fear. Under my skin, deep in my belly. And once I could see it, I couldn't *un*see it. It was nauseating. Fear of being othered, fear of losing housing, fear of loss of loved ones—

just to name a few. And I was so deeply tired.

Some fear is natural for any being, and helpful at times. To anticipate or respond to danger. We are *animals*. We eat, we drink, we sleep, we love, we play, we mate, we caress, we birth, we create, we sing, we shout, we fight, we nurture, we grieve, we suffer, we kill, we harvest, we live, we die.

I will die.

But this fear was on a whole other scale. It felt as though it had been woven into the fabric of our society and had infiltrated our lives, without us even seeing it. It had settled deep in my core.

The real world is enmeshed within an infinite number of relationships. We are dependent on people, dependent on the Earth. Dependent on the fungi that binds the roots to the soil that feeds the plants, which open their stoma to the air that we breathe into our lungs. And the insects feed on plants, as we do, and the bats and the birds eat the insects, flying through the air that we breathe. *Reconnect.*

Breathing, reconnection. The antidote to fear. We are supported by our communities and ecosystems. The seasons and cycles come and go. And we are still here, within it all. Part of it. In this moment.

Look up at the moon. *We are not alone.*

The sun and its seasons ground me into my local place, and offer a rhythm for life —times for rest, and times for doing on a daily and annual cycle. But the moon connects me to something bigger. When I look up at the moon, I feel rooted to the Earth, anchored to this nucleus around which the moon orbits.

I see the moon as a unifier, dancing between night and day, and any misconstrued dualities or separation. Unlike the seasons—where my distant family is in the depth of winter while we are lazing in the summer heat—no matter where we are on Earth, we can all watch the moon wax and wane at the same time in our calendar.

The moon's appearance rotates according to latitude (and time of day or night as it passes in an arch across the sky). The waxing crescent moon more-or-less appears to us as a 'c' here in the southern hemisphere, appears flipped in the northern hemisphere, and is a 'u' near the equator.[44] But we are all seeing the same moon phase, and even the same side of the moon.[45]

The moon offers a universality, holding us with *all* the beings. All the Earth. Even our ancestors too. All those that came before us, looked up at this very same moon.

The moon-ometer

Despite my best intentions to slow down, it's easy to get swept up in the rush of capitalism and modern life. I now use the moon as a marker or 'barometer' for earthly disconnection (a moon-ometer?). If I realise I haven't seen the moon in several days, or I don't know what phase it's in, it means I've lost touch with 'reality'.

It means I need to stop and connect into what is here, and now, and all around us. Because certainly, until recently, regardless of our place on Earth, the moon helped shape—and was an intrinsic part of—life. It is one of the strands in the ecological web. Even if it has been forgotten or lost to our awareness.

The moon is not just a celestial rock in our orbit, and it's not just a part of our ecology. It can also inspire mystery, awe, wisdom and respect. It can be viewed as an elder, welcoming the light, as well as the darkness. And the true beauty, in my view, is that the moon can be all these things; inviting us into our place in something bigger.

EPILOGUE

This web

Just as I closed my laptop after finishing writing 'Earthshine', my partner's pager goes off and he runs out the door to head to the brigade and onto the firetruck.

My phone beeps. It's not a bushfire alert, thankfully. It's a text message from one of the neighbours. 'We picked lots of mulberries this morning, would you like some?'

Barefoot, I go outside. The heat is oppressive. Heavy. And smoke fills the air. It's just a couple of weeks since bushfire shook the town, and I gently dissolve the fears that have arisen in my mind, remembering how much we've all been there for one another.

She passes the bowl of mulberries over the fence and they fill my white enamel colander.

Before I go back inside, I grab the steel watering can and head towards the water tank to fill it, so that I can top up the chickens' water. Startled from my haze, I freeze when I see something in the leaf litter. I smile. It's a Blue-tongue Lizard seeking shade and moisture on the east side of the tank, where the ground has seeped with water earlier in the day when I was filling watering cans and connecting hoses to the vegetable beds. So instead, I head to the tap on the side of the house, leaving 'Bluey' in peace.

Back inside, I portion the berries into containers, intermittently popping one or two in my mouth. The sweet, juicy tanginess is a delight on my tongue, and I close my eyes to enjoy each one. As I reach the bottom of the colander, I notice that my hands are stained a glorious reddish-purple.

This messy web is resilient, and is calling us into respect and reciprocity—simply to be a part of it.

NOTES

† A note on names (from p. 11)

The names of those dainty orchids and wildflowers felt perfect to me at the time. But of course, all these plants already had other names, unknown to me, for thousands of years.

Names are strange things to contemplate. They are like a signpost to a place, but not the place itself.

Common names of species are often styled in lower case. Though I feel that capitalising them helps give a level of importance and worthiness that they deserve. It's also helpful clarification; because is it a little brown bat, or a Little Brown Bat?

In this book, I followed this logic, capitalising the names of species, and using lower case for more generic labels.

But it's blurry-lined business. And sometimes I wonder about the focus on names at all.

Taxonomy has become increasingly important, naming and cataloguing the vast diversity of life on Earth. Though it's a relatively blunt tool when used for informing conservation—where species are often seen to have greater value than subspecies, populations, or individuals.

Names and language convey knowledge and meaning, however they also come with preconceived ideas. Unfortunate naming of some species can influence support for their protection, especially when names have negative associations—such as for Australian native 'rats'.[46]

I sometimes think about the way we might perceive individual beings based on their naming. Do we think of Crimson Rosellas *all* just as Crimson Rosellas, even though we think of humans and our pets so uniquely? Like the beloved cats Piper,

Daphne, Rook, Tiny, Rogue, Quinn—each with their own character?

In Ursula Le Guin's 'She unnames them', all the different animals agree to liberate themselves from the bounds of their naming.[47] The narrator in the story, who also chooses to become nameless, describes the immense closeness she feels with them after the unnaming.

Do names give a separateness? Does it stop the curiosity? Do we learn the name in the field guide, or tick them off the list, and then move on, instead of being in deeper relationship? To *know* them?

What if we discarded names all together, as Le Guin plays with? We'd also be discarding a shared vocabulary, which is not at all practical. But would it open something up? Maybe we'd all then be placed in simultaneous importance and lack of importance. To be? And to be *with*?

While names are likely here to stay, it's an interesting thought-experiment. And when a Crimson Rosella flashes past, or a 'Maggie' softly warbles on a branch overhead, I try to take notice—and to stop and listen to what they might have to say.

‡ A note on caves (from p. 17)

Bats are vulnerable to human disturbance. Please help protect them by not entering caves that are used by bats. For the field research I referred to in 'Learning from bats', I had a range of permits and animal ethics approvals, and adhered to strict protocols to minimise disturbance. The cave that I referred to in 'Earthshine' was not a bat cave.

Human presence in roosts can cause bats to 'wake' from torpor, which might waste valuable fat stores and compromise survival, or lead to site abandonment.

People can also inadvertently spread diseases like White-nose Syndrome, which can spread by fungal spores on boots, clothing and equipment and has caused mass population declines of bats.

Acknowledgements

It doesn't just take a village, it takes a whole ecosystem. To all, I am very grateful. In particular, I want to thank my first readers (and editors) Joe, Jessie and Riva; Frida, for transforming my photo into a beautiful cover design, and for converting my hand-drawn illustrations into digital images for the book; and my parents, for always supporting me.

About the author

Dr Emmi van Harten lives on Djaara country with her family, chickens and inside cats. Her parents and siblings emigrated to Australia from the Netherlands before she was born. In a world that often felt nonsensical, Emmi turned to nature to find her place and make sense of things. She is an ecologist specialising in bats, and her latest love (of many) is cycling her bike Eleanor along the local creek.

References cited

1. Reed EH, Bourne SJ. 2000. Pleistocene fossil vertebrate sites of the south-east region of South Australia. *Transactions of the Royal Society of South Australia.* 124(2):61–90.

2. Codd J, Clark B, Sanderson K. 1999. Drinking by the Common Bent-wing Bat *Miniopterus schreibersii* and calcium in cave water. *Bat Research News.* 40(1):9–10.

3. Kwaymullina A. 2020. *Living on Stolen Land.* Magabala Books: Broome.

4. Villada-Cadavid T, Wu NC, Sloggett B, Lumsden LF, Welbergen JA, Turbill C. 2025. Winter torpor and body mass patterns of a cave-roosting bat in cool and warm climates. *Oecologia.* 207(12):193. doi:10.1007/s00442-025-05835-9

5. Crichton EG, Seamark RF, Krutzsch PH. 1989. The status of the corpus luteum

during pregnancy in *Miniopterus schreibersii* (Chiroptera: Vespertilionidae) with emphasis on its role in developmental delay. *Cell and Tissue Research*. 258(1):183–201. doi:10.1007/BF00223157

6. May K. 2021. *Wintering: The Power of Rest and Retreat in Difficult Times*. Rider: London.

7. Wilson J. 2003. *Flowering Ecology of a Box-Ironbark Eucalyptus Community*. PhD thesis. Deakin University.

8. Soderquist T, Rhind S. 2023. Brush-tailed Phascogale. In: Baker AM, Gynther IC, eds. *Strahan's Mammals of Australia*. Fourth edition. Reed New Holland Publishers: Wahroonga. 108–110.

9. Hersey T. 2022. *Rest is Resistance: Free Yourself from Grind Culture and Reclaim Your Life*. Aster: London.

10. Geiser F, Bondarenco A, Currie SE, Doty AC, Körtner G, Law BS, Pavey CR, Riek A, Stawski C, Turbill C, Willis CKR, Brigham RM. 2021. Hibernation and daily torpor in

Australian and New Zealand bats: does the climate zone matter? *Australian Journal of Zoology*. 67(6):316–330. doi:10.1071/ZO20025

11. Payne C. 2024. Oh, miserable modernity! with Yin Paradies. *Reskillience* [Podcast].

12. van Harten E, Lawrence R, Lumsden LF, Reardon T, Bennett AF, Prowse TAA. 2022. Seasonal population dynamics and movement patterns of a critically endangered, cave-dwelling bat. *Wildlife Research*. 49(7):646–658. doi:10.1071/WR21088

13. Baudinette R, Wells R, Sanderson K, Clark B. 1994. Microclimatic conditions in maternity caves of the bent-wing bat, *Miniopterus schreibersii*: an attempted restoration of a former maternity site. *Wildlife Research*. 21(6):607–619. doi:10.1071/WR9940607

14. Bail NJ, Lumsden LF, Reardon T, van Harten E, Clissold P, Prowse TAA. 2025. Emergence and return times in a colonial, cave-dwelling bat: age and sex differences driven by reproductive cycle. *Ecology and*

Evolution. 15(5):e71281. doi:10.1002/ece3.71281

15. Morton SR. 2017. On pessimism in Australian ecology. *Austral Ecology*. 42(2):122–131. doi:10.1111/aec.12410

16. Penman S. 2019. Prof Lesley Hughes. *Terrain: Stories of Australian Women in Science* [Podcast]

17. Park A, Williams E, Zurba M. 2020. Understanding hope and what it means for the future of conservation. *Biological Conservation*. 244:108507. doi:10.1016/j.biocon.2020.108507

18. Darabont, F. 1994. *The Shawshank Redemption* [Film]. Columbia Pictures.

19. Nagel T. 1974. What is it like to be a bat? *The Philosophical Review*. 83:435–450.

20. NASA. 2026. Moon phases. Available at: https://science.nasa.gov/moon/moon-phases/

21. NASA. 2026. Tides. Available at: https://science.nasa.gov/moon/tides/

22. Wilkinson K, Palmer G. 2022. Ambient light energy intensity as a trigger for the dawn chorus: patterns in five common eastern Australian bird species. *Australian Field Ornithology*. 39:82–88. doi:10.20938/afo39082088

23. Grant R, Halliday T, Chadwick E. 2013. Amphibians' response to the lunar synodic cycle—a review of current knowledge, recommendations, and implications for conservation. *Behavioral Ecology*. 24(1):53–62. doi:10.1093/beheco/ars135

24. Saldaña-Vázquez RA, Munguía-Rosas MA. 2013. Lunar phobia in bats and its ecological correlates: a meta-analysis. *Mammalian Biology*. 78(3):216–219. doi:10.1016/j.mambio.2012.08.004

25. Roeleke M, Teige T, Hoffmeister U, Klingler F, Voigt CC. 2018. Aerial-hawking bats adjust their use of space to the lunar cycle. *Movement Ecology*. 6(1):11. doi:10.1186/s40462-018-0131-7

26. Bischof R, Vallejo-Vargas AF, Semper-Pascual A, Schowanek SD, Beaudrot L, Turek D, Jansen PA, Rovero F, Johnson SE,

Guimarães Moreira Lima M, Santos F, Uzabaho E, Espinosa S, Ahumada JA, Bitariho R, Salvador J, Mugerwa B, Sainge MN, Sheil D. 2024. The moon's influence on the activity of tropical forest mammals. *Proceedings of the Royal Society B.* 291(2033):20240683. doi:10.1098/rspb.2024.0683

27. Prugh LR, Golden CD. 2014. Does moonlight increase predation risk? Meta-analysis reveals divergent responses of nocturnal mammals to lunar cycles. *Journal of Animal Ecology.* 83(2):504–514. doi:10.1111/1365-2656.12148

28. San-Jose LM, Séchaud R, Schalcher K, Judes C, Questiaux A, Oliveira-Xavier A, Gémard C, Almasi B, Béziers P, Kelber A, Amar A, Roulin A. 2019. Differential fitness effects of moonlight on plumage colour morphs in Barn Owls. *Nature Ecology and Evolution.* 3(9):1331–1340. doi:10.1038/s41559-019-0967-2

29. Kronfeld-Schor N, Dominoni D, De La Iglesia H, Levy O, Herzog ED, Dayan T, Helfrich-Forster C. 2013. Chronobiology by moonlight. *Proceedings of the Royal*

Society B. 280(1765):20123088. doi:10.1098/rspb.2012.3088

30. Gao B, Hu G, Chapman JW. 2024. Effects of nocturnal celestial illumination on high-flying migrant insects. *Philosophical Transactions of the Royal Society B.* 379(1904):20230115. doi:10.1098/rstb.2023.0115

31. Storms M, Jakhar A, Mitesser O, Jechow A, Hölker F, Degen T, Hovestadt T, Degen J. 2022. The rising moon promotes mate finding in moths. *Communications Biology.* 5(1):393. doi:10.1038/s42003-022-03331-x

32. Dacke M, Nilsson DE, Scholtz CH, Byrne M, Warrant EJ. 2003. Insect orientation to polarized moonlight. *Nature.* 424(6944):33. doi:10.1038/424033a

33. Freas CA, Narenda A, Murray T, Cheng K. 2024. Polarised moonlight guides nocturnal bull ants home. *eLife.* 13:RP97615. doi:10.7554/eLife.97615

34. Sannidhi S, Priyanka G, Singiri JR, Novoplansky N, Grafi G. 2026. Plants and the moonlight: a controversial subject

revisited. *Plant Science*. 362:112841. doi:10.1016/j.plantsci.2025.112841

35. van Hasselt SJ, Rusche M, Vyssotski AL, Verhulst S, Rattenborg NC, Meerlo P. 2020. Sleep time in the European Starling is strongly affected by night length and moon phase. *Current Biology*. 30(9):1664-1671.e2. doi:10.1016/j.cub.2020.02.052

36. NASA. 2026. Moon facts. Available at: https://science.nasa.gov/moon/facts/

37. McNamara JA [translator]. 1997. The Life of St. Eligius, 588–660. In: Halsall P, ed. *Medieval Sourcebook*. Available at: https://sourcebooks.fordham.edu/basis/eligius.asp

38. Cajochen C, Altanay-Ekici S, Münch M, Frey S, Knoblauch V, Wirz-Justice A. 2013. Evidence that the lunar cycle influences human sleep. *Current Biology*. 23(15):1485–1488. doi:10.1016/j.cub.2013.06.029

39. Casiraghi L, Spiousas I, Dunster GP, McGlothlen K, Fernández-Duque E, Valeggia C, De La Iglesia HO. 2021. Moonstruck sleep: synchronization of

human sleep with the moon cycle under field conditions. *Science Advances.* 7(5):eabe0465. doi:10.1126/sciadv.abe0465

40. Wehr TA. 1992. In short photoperiods, human sleep is biphasic. *Journal of Sleep Research.* 1(2):103–107. doi:10.1111/j.1365-2869.1992.tb00019.x

41. Ekirch AR. 2001. Sleep we have lost: pre-industrial slumber in the British Isles. *The American Historical Review.* 106(2):343–386. doi:10.2307/2651611

42. Windred DP, Burns AC, Rutter MK, Lane JM, Saxena R, Scheer FAJL, Cain SW, Phillips AJK. 2025. Light exposure at night and cardiovascular disease incidence. *JAMA Network Open.* 8(10):e2539031. doi:10.1001/jamanetworkopen.2025.39031

43. Deprato A, Maidstone R, Cros AP, Adan A, Haldar P, Harding BN, Lacy P, Melenka L, Moitra S, Navarro JF, Kogevinas M, Durrington HJ, Moitra S. 2024. Influence of light at night on allergic diseases: a systematic review and meta-analysis. *BMC Medicine.* 22(1):67. doi:10.1186/s12916-024-03291-5

44. Howells K. 2020. Can the moon be upside down? The Planetary Society. Available at: https://www.planetary.org/articles/can-the-moon-be-upside-down

45. NASA. 2026. Tidal locking. Available at: https://science.nasa.gov/moon/tidal-locking/

46. Morton S. 2024. Why don't people care about Australia's native rodents? The problem could be their ugly names. *The Conversation.* 1 February. doi:10.64628/AA.srvkg46gv

47. Le Guin UK. 1985. She unnames them. *The New Yorker*. 21 January. Available at: https://www.newyorker.com/magazine/1985/01/21/she-unnames-them

www.ingramcontent.com/pod-product-compliance
Lightning Source LLC
LaVergne TN
LVHW040516210926
848480LV00005BA/1476

* 9 7 8 0 6 4 6 7 3 8 5 2 9 *